333.72
J   Jones, Claire
        Pollution: the land we live on ⸢by⸣ Claire
    Jones, Steve J. Gadler, ⸢and⸣ Paul H. Engstrom.
    Minneapolis, Lerner, ⸢1971⸣
      102p.   illus.   23cm.   (Real World book)

       1.Pollution. 2.Conservation of natural
    resources. I.Gadler, Steve J., jt. auth.
    II.Engstrom, Paul H., jt. auth. III.Title.

POLLUTION:

the LAND
we live on

**These books are printed on
paper containing recycled fiber.**

a real world book

POLLUTION:

the LAND
we live on

Claire Jones
Steve J. Gadler
Paul H. Engstrom

LERNER PUBLICATIONS COMPANY
Minneapolis, Minnesota

Acknowledgments

The illustrations are reproduced through the courtesy of: p. 6, Hilmar Pabel; p. 10 (top and bottom), Henry Valinkas; pp. 18, 25, 33, 39, 51, 84, 93, United States Department of Agriculture; p. 22, United States Department of the Interior, Bureau of Reclamation; pp. 23, 73, United States Forest Service; p. 28, Michigan Department of Natural Resources; pp. 32, 57, 78, 79, 81, Environmental Protection Agency; p. 37, United Press International, Inc.; p. 43, United States Steel; p. 46, United States Atomic Energy Commission; p. 49, Citizen's Advisory Committee on Environmental Quality; p. 55, The American Institute of Architects; p. 59, Caterpiller Tractor Co.; p. 60, Grant Heilman; p. 68, The Black Clawson Company; p. 70, Reynolds Metals Company; p. 74, United States Department of the Interior, National Park Service; p. 89, Kaiser Center, Inc.; p. 91 (all photos), Ian L. McHarg, Narendra Juneja and Lindsay Robertson in "A Comprehensive Highway Route Selection Method, Applied to I-95, New Jersey" by Wallace, McHarg, Roberts and Todd 1965.

International Standard Book Number: 0-8225-0629-7
Library of Congress Catalog Card Number: 74-156365

Second Printing 1972

Contents

This gigantic wheel excavator is a modern strip-mining machine. It can chew up hundreds of thousands of tons of soil and turn a landscape into a coal mine within a day.

1

Ravaging the Land

In Belmont County, Ohio, a monstrous machine, 100 feet high, moves night and day over the green land, tearing out giant chasms and making the earth as bleak as the moon. The people of tiny Kirkwood listen to the machine's ear-splitting howl as it completely wipes out much of their town. Romantically misnamed the "Gem of Egypt," this man-made monster is strip-mining for coal and destroying what was once productive farm country. While the coal it harvests goes to electric power companies in the Midwest, the machine itself consumes more electricity in its work than any city in Belmont County.

In Minnesota, much of a 500-acre freshwater lake in the Mississippi River valley was destroyed because the Port Authority of St. Paul decided to turn it into an industrial park. Called Pig's Eye Lake, after the nickname of an early French trader, this beautiful area of water and marshland once provided food and shelter for deer, beaver, muskrat, and many other wild creatures. It is on a major continental flyway for migrating birds. But in 1968 filling began. A steel factory, fertilizer plant, and barge terminal have been built there. A network of high tension power lines now crisscrosses the area. Much of the lake had been lost to industry before outraged citizens were able to stop the work of destruction in 1969. What remains has been preserved, even though its value to wildlife is much reduced.

There was jubilation in Denver, Colorado, when the International Olympics Committee decided to hold the 1976 Winter Olympics there. But once plans for this great event were drawn up, citizens of the small town of Evergreen up in the mountains nearby were less than delighted. Acres of their green land are to be paved over for parking lots, 25 acres of trees will be felled for the bobsled run, and an eight-foot-wide swathe is to be chopped out of the forest for a distance of 55 miles for a cross-country ski trail. To make matters worse, Evergreen has no zoning laws. Real estate developers can move in with enterprises designed to make a quick profit from the crowds expected for the Olympics and then pull out— leaving behind great scars upon the land.

These attacks upon the land in Ohio, Minnesota, and Colorado are only three examples of how we ravage and destroy our mother earth. We bury our land under a million tons of garbage every day. We wipe out whole species of plants and animals with chemical pesticides and herbicides. We cut down trees because they get in the way of our billboards, or because it is less trouble than building a sidewalk around them. We turn landscapes of startling natural beauty into disaster areas of neon signs and hamburger stands.

This land the white man took from the Indian was glorious in its wild and varied beauty. Even two centuries ago, man-made scars were scarcely visible on its soaring mountains, whispering deserts, and rolling plains. The swamps teemed with wildlife. Jungles, forests, hills, and lakes stood as they had for thousands of years. The wild rivers plunged freely. But, in the years between, modern man has wreaked havoc with his lust for prosperity and his scorn for untamed nature. Wherever he has congregated, America the beautiful has rapidly become a slum.

Many thousands of years ago when man was emerging from primitive ancestry, he was just one small part of the natural environment as were all the other animals. In many respects, he was poorly equipped for survival. He moved slowly on his feet, lacked effective weapons of tooth and claw, had relatively poor eyesight and sense of smell, and grew no heavy coat of fur to help him through cold winters. But he began to use his brain to control the environment to make life safer and more comfortable.

Where once he lived on wild roots, berries, and nuts, man began to plant and harvest crops. He chipped out weapons to hunt more effectively and to protect himself from predators. He gathered into communities, sharing the tasks of survival. When life became more secure, man developed more complex civilizations—irrigating the land for more intensive agriculture, building homes and cities, chopping down trees, and plowing up grasslands. He learned to manufacture goods for himself and for trading. He began to gouge out the land for minerals to make these goods.

Two views of the same area along the Mississippi River in Minnesota. In 1910 (above) the land was part of a 2,500-acre park,
but by 1969 (below) industry had been allowed to totally ravage
the riverfront.

Nevertheless, for hundreds of thousands of years, human beings lived within the worldwide *ecosystem*—the natural system of relationships among all living things and their environment. On occasions when we upset the balance of nature in our own small environment, there was room to move away and start again somewhere else. But as we prospered and our communities grew, we no longer wanted to keep moving on. We learned even more how to adapt our environment to our needs and, gradually, how to dominate the natural ecosystem.

Now, we talk of conquering nature, of harnessing nature to serve us better. We see ourselves as the center of the universe and the reason for its very existence. We cultivate or exterminate other living things without agonies of guilt. Our technology has developed to the point where we can and do move mountains, reroute rivers, and blot out lakes. When we upset the balance of nature today, we do it on a worldwide scale.

Most of the time we don't even understand the damage we are doing to the natural ecosystems of the world. Nor do we understand how this damage is beginning to threaten our own survival. Mostly we have not even tried to understand because in many parts of the world our cultures have become so commercialized that we put our best efforts only into knowledge and action that bring us profit and prosperity.

We, of the species man, are the highest forms of life on earth. It is essential for our survival that we learn to use our intelligence and technology to live within our niche in the balance of nature. To do this we need to understand very much more than we do now about the complicated relationships within the natural life cycles.

2

The Cradle of Life

How the Land was Formed

About 5 billion years ago, our planet had its beginning as a hot, gaseous sphere. After millions of years, its hot mass had cooled to a solid. Then, scientists think, more millions of years passed while our earth alternately melted and cooled and its atmosphere slowly developed. Not until about 3.5 billion years ago had our planet formed its final solid crust.

Time on this scale is nearly impossible to imagine. But we begin to get some idea if we convert these figures into a scale we can comprehend. Assume that the earth was a gaseous ball one year ago. It would then have cooled to the point where it had a solid crust and an atmosphere about nine months ago. One month ago the earliest forms of plant life on land were developing. Man as a species has existed only for 3.5 hours. Jesus was born 12 seconds ago, and Columbus discovered America just about 3 seconds ago. The modern developments that have shaped our lives happened in the last second of the year. This gives some idea of how long our planet existed before man became a dominant force.

Even after the earth's crust had formed, and the low places of the crust had filled with water and been transformed into ocean beds, the surface of our earth was not stable. The land masses and ocean floors kept on changing their shapes—as they continue to do. Sometimes the changes occurred with the sudden violence of earthquakes and volcanic eruptions. But they also took place more slowly through vast pressures from within the earth squeezing and forcing its crust into altered shapes.

The chemistry of the newly formed world was not stable either. Elements from the atmosphere combined with elements from earth's rocks. The wind and the rain slowly eroded the rocks and washed their minerals into the oceans. Over many millions of years, lightning and the sun's radiation acted upon the new compounds which had formed in the atmosphere and in the seas to create the chemistry from which life began.

But even after life had appeared in the seas, our planet had to undergo many other changes before life could exist on the land. One of these changes was the slow formation of soil from the earth's rocky surface.

The forces of *weathering* were primarily responsible for breaking down rocks into soil—just as they are today. For example, air contains elements (such as oxygen) which combine with elements in rocks, causing the rocks to "decay" into smaller particles. Wind, rain, and running water all break down rocks physically, by simply wearing

13

them away, and also chemically, by combining some of their elements with elements in the rocks. Ice too helps to create soil because water lying in cracks in rocks expands when it freezes and acts as a powerful wedge to break the rocks apart.

However, soil formed only by weathering, really rock dust, does not contain enough food to support plant life. To support plants, soil must contain ingredients which are supplied by the decay of once-living things. The natural formation of a single inch of such fertile soil, *topsoil*, today requires between 300 and 1,000 years. But how was our earth's first topsoil formed?

Obviously, the first simple plants living on the land could not need fertile soil to exist: there was none. But when they died, their decayed remains mixed with small particles of rock dust to start a soil fertile enough to support the higher forms of plant life. Lichens, a primitive form of plants existing today, show us how the formation of topsoil by the earth's first plants may have begun. Made up of an alga and a fungus working together, lichens live on bare rocks and actually break down the rocks they live on with acids they secrete. When lichens die, they add their decayed matter to the rock dust. When there is enough of a mixture of soil and decayed matter, plants which need fertile soil are able to replace some of the lichens. The roots of these plants help to break down the rocks even further, and the plants also enrich the soil with decayed matter when they die. Eventually, there is enough topsoil to support larger plants, and then shrubs and trees.

The formation of a surface of fertile soil was essential for the development of plant life on the land. Only after plants had become established on the earth and worked their own chemistry on the atmosphere and the land was the environment able to support animal life.

Land and the Ecosystem

The land we live on is mother to us all—not just to every human being but also to all the land-based animals, plants, and microorganisms which make up a delicately balanced ecosystem of which we are just one part. Ecosystem is the scientific term for all the living and non-living things in a given area and the relationships among them. Our entire world is an ecosystem, and so is a continent like North America. But a small area of forest or a pond in an open field are also complete ecosystems in themselves.

Within any ecosystem, large or small, one of the most important relationships involves how food and energy move through the system in what is called a *food chain*. Ecologists divide an ecosystem into four main parts which roughly describe the relationships among the members of a food chain: nonliving substances, *producers, consumers,* and *decomposers.*

Green plants are producers which take nonliving substances such as minerals and water from the soil and carbon dioxide from the atmosphere. They use these substances and convert them into oxygen and foods which animals use to make living tissues. We and the other animals are consumers: we eat plants and other animals. (We also take in oxygen with the air we breathe, convert it into carbon dioxide, and exhale it into the atmosphere where it becomes available for plants to use.) Decomposers such as fungi and bacteria feed on animal excretion and

15

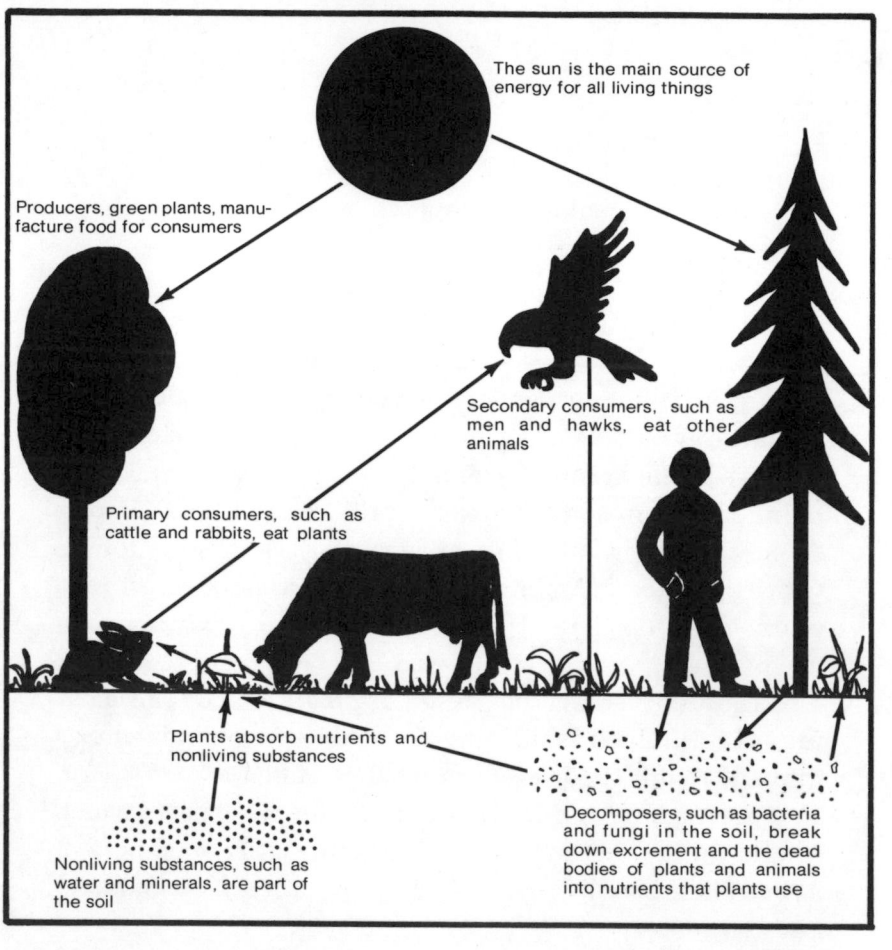

Diagram of a simplified ecosystem showing the main parts of a food chain. If any link in the chain is altered or destroyed, the entire ecosystem will be affected.

dead plants and animals, breaking them down into nitrogen, minerals, carbon dioxide, and water. They release such substances back into the ecosystem, making them available to be used again by plants. Without the activities of its decomposers, our earth would be buried beneath putrid dead animals and plants, and important substances such as nitrogen would not be recycled through the ecosystem.

Topsoil is one of the most intricate works of nature and an all-important part of any land-based ecosystem. It is the home and principal feeding ground of the bacteria, fungi, and higher plants (decomposers and producers), as well as worms, insects, and other small animals. The topsoil of our land contains the mineral nutrients and other nonliving substances which plants extract, and use, and pass on up the food chain to us and the other animals.

Nitrogen, for example, which is an important element for the growth of living tissues, enters the food chains of the world through the soil. About 78 percent of the earth's atmosphere is made up of free nitrogen, but animals and plants are not able to use it directly out of the air. Before nitrogen can enter the world's life cycles, it has to be converted into salts by the few living organisms which have the ability to use free nitrogen. The most important of these are the nitrogen-fixing bacteria which live on the roots of clover, alfalfa, peas, beans, and other plants in the legume family. Once these bacteria have converted free nitrogen into salts in the soil, plants are able to absorb the salts through their roots and so pass them up the food chain to all living creatures. Nitrogen is returned to the soil and to the air by animal excretions and by the death and decay of plants and animals. Decomposers work upon the dead organic matter and release the nitrogen again for recycling through the soil.

The chemical and physical structure of the topsoil varies from place to place, depending upon many things, such as the natural minerals in the land, the amount of water available, and the temperature of an area. And the kind and number of plants and animals that can live in an area depend very much on the nature of its topsoil. For

A profile of fertile land showing nearly two feet of dark topsoil above the lighter layer of subsoil. The kind of life that any land can support depends upon the depth and quality of its topsoil.

example, only a few of the decomposers which are so important for soil fertility can live and work in very cold areas.

In areas where the soil is moist and contains humus, earthworms can flourish and work on the earth. The earthworm has long been an unsung hero of the land. He feeds on dead leaves and other organic matter, breaking it down as it passes through his digestive tract, and then enriching the soil when he excretes it. By constantly working his way through the soil, he aerates it, supplying air for bacteria to use in their work of converting nitrogen into salts and of decomposing dead matter. His tunneling also helps water to diffuse through the soil and makes it easier for the roots of plants to penetrate the soil. Perhaps his most important contribution is actually building up the layer of topsoil. He is continuously depositing soil from the layer of *subsoil* below ground onto the surface of the earth. Charles Darwin, who shaped the present-day theory of evolution, once calculated that earthworms can add a layer of topsoil one to one-and-a-half inches thick in 10 years.

Animals that live on the surface of the land also affect the quality of its soil. They eat the plant cover, sometimes stripping it so bare that there is nothing to help the soil hold moisture or even to hold the topsoil in place. Animals also improve the land, however, by burrowing and digging in it and by giving it their excrement and dead bodies.

We, too, affect the soil. Primitive men worked on the land in the same way as the other animals. But now that we cultivate it so widely, we affect it in more drastic ways. We change the vegetation and the animal life. We remove minerals from the land by intensive agriculture and by mining. We cover it with fertilizers, pesticides, and herbicides. We drain it and irrigate it. We do all this continuously, often without realizing the effects of our actions on the intricate balance of nature.

3

The Damage Man is Doing

Agriculture

Our early nomadic ancestors were not concerned with land conservation. They farmed the same land season after season until it had grown too poor to yield the food they needed. Then they moved on to plant their crops somewhere else. Depending upon the amount of rainfall in the area, the average temperatures, the depth of topsoil, and the amount and kinds of vegetation that remained, the exhausted and abandoned land might take years, decades, or even centuries to recover.

Some of the greatest civilizations of the past went into decline because they abused their land. During his travels across Europe and Asia in the thirteenth century, Marco Polo saw Mesopotamia (comprised of parts of present-day Iraq, Syria, and Turkey) as a rich and fertile land of green. It supported a large population with a farming economy based on *irrigation*. Eventually the agricultural system broke down, however, partly because the irrigation ditches filled with silt and partly because the land became overloaded with mineral salts. Much of what was called Mesopotamia is desert today.

All water accumulates dissolved minerals as it filters through the land and runs over the beds of streams and rivers. In hot, dry areas, when water is channeled into irrigation ditches and made to trickle out over the land, some of it evaporates in the sun and leaves its minerals in the form of salts deposited on the land. A certain amount of mineral salts can be beneficial to crops. But too much is fatal. As the processes of irrigation and evaporation go on, the land becomes too overloaded with mineral salts for plants to grow at all. For all his patient care, man finds he has created a desert. And because this happens in hot, dry areas with little natural rainfall to flush the salts away, the desert usually remains barren for thousands of years.

At the present time, the Imperial Valley in southern California is in danger of destruction by salt deposits. By diverting the waters of the Colorado River and providing irrigation for every field, man has made this sun-baked plain blossom into one of the most productive farming areas in the world. The Imperial Valley can support two, three, and even more crops each year on the same land. But the waters of the Colorado River carry huge quantities of mineral salts by the time they reach southern California. After years of irrigation and evaporation, these salts are building up in the farmland to the extent that some growers have already had to switch from vegetables to more salt-resistant, and less profitable, crops. Unless some way can be found to remove the salts from the water or from the land, the Imperial Valley will revert to desert. But this time it will be a desert not reclaimable for farmland.

There are many ways in which we thoughtlessly exhaust or destroy our land. For instance, it is estimated that 50 million acres of land in the United States have been destroyed by soil *erosion*.

Ground cover of shrubs, trees, and other plants protects the land by helping it to absorb rainfall and by sheltering it from fierce winds. Whenever this cover is removed by intensive grazing, by farming, by timbering, or by any-

The Coachella Canal in the Imperial Valley, California, has enabled man to create gardens in the desert. But if salt deposits caused by irrigation are allowed to build up in the soil, fields and groves will revert to unreclaimable desert.

A tree has protected the soil around it on this otherwise heavily eroded California hillside where overgrazing stripped the land of its protective cover.

thing else, the land is in danger of erosion. Rain washing over the land surface will carry loose topsoil away with it. Much of the topsoil is deposited in streams and rivers, and some of it is carried by the water out into the oceans, where it can no longer enrich the land. The life-supporting organic matter of the topsoil is lighter in weight than other parts of the land and therefore more easily removed. Once it is gone, fertility is destroyed and the land no longer supports protective vegetation. It remains bare and subject to ever more erosion by the wind and the rain.

On sloping hillsides the danger of erosion is even greater because water moves more swiftly over the surface. Some farming communities on steeply sloping land are able to survive by contour plowing—plowing furrows across rather than up and down a hill. The furrows then catch the rainwater and prevent it from running down the hill. Some farmers use stair-step terraces so that at each contour of a hill, the land is level. Others are careful to plant protective covering wherever they can.

Trees are essential for a healthy environment on steeply sloping lands. Their long roots reach down into the soil, helping it to absorb rain. The network of roots also helps to keep the soil in place. When trees are stripped off or destroyed by forest fires, mountainous lands can quickly lose their topsoil and revert to bare rock. The slow cycle of breaking down rock into soil will begin again, but it takes thousands of years for new soil to become productive. Erosion can wash or blow away productive soil in a few years, or even weeks.

In *The Grapes of Wrath*, John Steinbeck wrote the tragic story of farmers and their families forced to leave the dustbowls of Oklahoma and Kansas in the 1930s. Their land had been destroyed in less than 50 years by farming methods which allowed the fertile topsoil to erode away in the wind of the dry plains.

During thousands of years while they were barely touched by man, the dry grasslands of the West became wonderfully fertile. Over most of the earth the depth of topsoil can be measured in inches. But on the western grasslands the low rainfall, which is not enough to wash mineral nutrients from the soil, combined with a rich growth of grasses to build a fertile topsoil in some places as deep as 20 feet. Man's first attempts at agriculture in the area were exceptionally rewarding. But his crops used the nutrients of the soil, and he took no care to replace them. The plants which grew there naturally had enriched the soil with their own decay. Their roots had kept the soil in place and helped it to hold water. Domestic crops were no substitute for the native grasses. They could not survive the periodic droughts. They died, and left the soil bare. Overgrazing of livestock also stripped the natural protective cover from the land. Then the winds blew and swept the topsoil away.

As the population of the world increases and more food is needed to feed us all, there are constant pressures on farmers to cultivate land unsuitable for agriculture and to increase the yield from all farmlands. Pressures come from poor countries with extreme needs and from wealthy

In this area of Nebraska where grasses no longer protect the land, wind has severely eroded the soil, leaving a sandy desert.

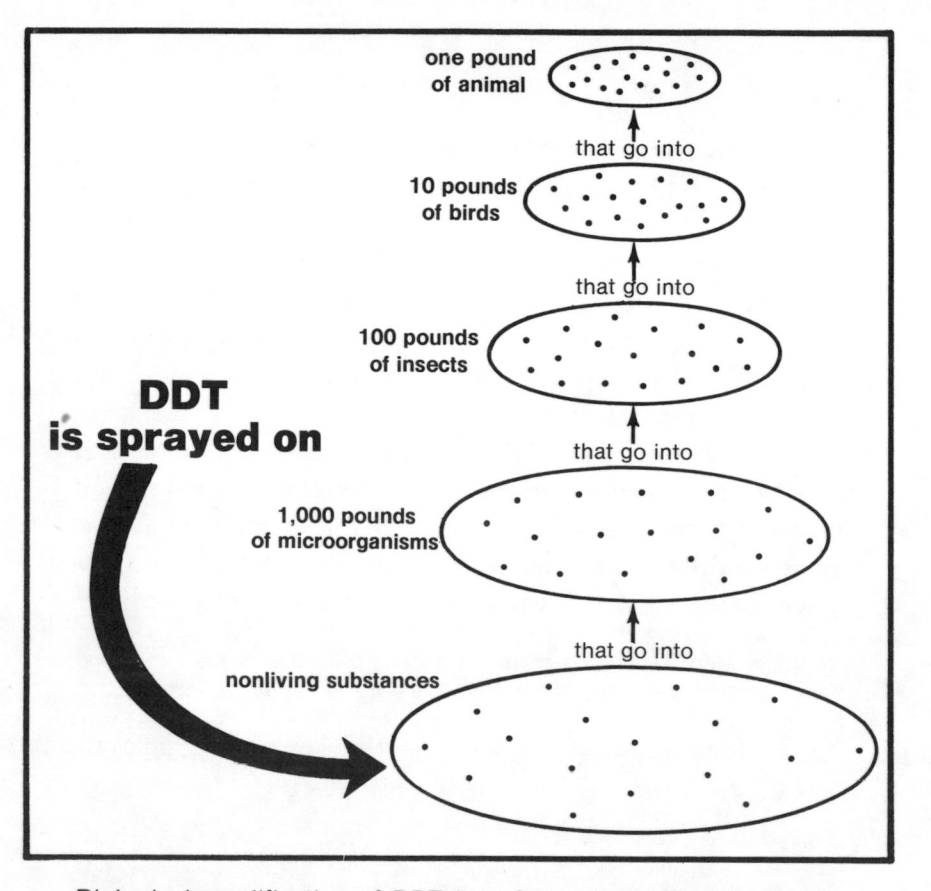

one pound
of animal

that go into

10 pounds
of birds

that go into

100 pounds
of insects

**DDT
is sprayed on**

that go into

1,000 pounds
of microorganisms

that go into

nonliving substances

Biological amplification of DDT in a food chain. (The black dots represent DDT.) When certain stable pesticides, such as DDT, are sprayed on soil or water, microorganisms absorb them and pass them up a food chain. And, since living things eat many times their own body weight in food, animals higher in a food chain accumulate larger amounts of pesticide than animals below them. At any level of a food chain, concentrations of pesticide may reach toxic amounts.

countries with the desire to get richer quicker. As a result, crops are protected from disease with pesticides. Weeds are destroyed with herbicides. Seeds are dipped in fungicides and insecticides. Fertilizers are used in increasing quantities. In the United States alone, we douse the land with millions of tons of virulent chemicals each year.

The pesticide DDT has been used generously through-out the world. It has increased the yield of crops by controlling plant diseases. It has protected many people against pest-borne diseases, including typhus, encepha-litis, bubonic plague, and cholera. But it is also having unexpected and devastating effects on the natural life cycles of the world, in hundreds of different ways.

DDT is an extremely stable chemical with a very long life, and its residues accumulate in the various food chains of the world by *biological amplification.* This means that DDT is passed from one living thing to another in a food chain and that each organism in the chain acquires a larger amount of DDT in relation to its body weight than does the organism below it. For example, microscopic bacteria feeding on soil which has been doused with DDT may accumulate perhaps 20 parts of DDT per million parts of themselves. Bugs which feed on the bacteria may absorb 1,000 parts of DDT per million parts of themselves. Birds that eat the bugs retain an even larger amount in relation to their body weight. And animals that eat the birds retain even more.

We already know that DDT can damage the ability of birds to use calcium so that the shells of their eggs are too thin to avoid breakage in the nest before the young birds hatch. Because of this, predatory birds, such as the bald eagle and the peregrine falcon, which accumulate higher concentrations of DDT than birds below them in the food chain, are in danger of extinction. There is also evidence that DDT causes sterility in creatures so that they lose the ability to reproduce themselves. Scientists believe in addition that sufficient accumulations of DDT contribute to the development of cancer. In the United States, concentrations of DDT in the milk of some human mothers now exceeds the safe levels for marketable foods set by the Food and Drug Administration.

As long ago as 1962, Rachel Carson drew popular attention to the dangers of DDT and other chemicals used in agriculture in her book *Silent Spring*. But only now are bans on some of these chemicals beginning to come into effect. Because of economic pressures, the relative cheapness of DDT, and determined disbelief about its worldwide dangers, DDT has been used long after its dangers were known.

Because of the effects of DDT, the eagle is now in danger of extinction. Accumulations of DDT in the mother eagle may be responsible for the egg that didn't hatch in this nest.

Much experimental work is being done today to develop safer ways of controlling pests. Already, there are some 65 different pesticides which can do a more selective job than DDT, with fewer harmful worldwide effects. Some experiments with these pesticides, however, are revealing new dangers.

Parathion, for instance, which is said to be a sure-fire killer of the worms which riddle some crops, also kills people. A tobacco grower in North Carolina used parathion on his crop after he learned that government regulations forbade him to use DDT if he wanted to qualify for price supports. After the recommended five-day waiting period, his seven-year-old son, Daniel, spent a long day in the field, then ate supper and went to bed. During the hot, humid night, he was cold and shivering. In the morning he was dead. Five days later his brother, Curtis, fainted in the barn and was rushed to a hospital where his life was saved by injections of an antidote.

Workers on California's fruit and vegetable farms have become nauseated and have gone into convulsions after contact with parathion. Six of nine cases of pesticide poisoning in one area of Florida were traced to parathion. Parathion is not a stable chemical like DDT; it breaks down so fast that there is no danger of it accumulating in the world's food chains as DDT does. But parathion is so lethal that it should be kept under lock and key and used only under tightly controlled conditions.

Other dangers are created when chemical pesticides are used in unnecessarily large quantities—like using a sledgehammer to crack a nut. The United States Department of Agriculture's fire ant program is one example. The fire ant is a pesky creature which builds nests in hayfields and bites field workers when it is disturbed. The bites are fiery and painful, but they are not hazardous to health. Nor is the ant any threat to agriculture. It feeds mainly on other insects, including the destructive boll weevil, the bark beetle, the pinetip moth, and the sugarcane borer. The ants could easily be controlled by doses of short-lived pesticides on each of their nests. Instead, the plan calls for a $200-million campaign over 12 years to wipe out this insignificant pest by blanketing 120 million acres of land in the South with 450 million pounds of bait laced with deadly Mirex pesticide. An aerial bombardment will spread the bait over every square foot of land in cities and fields. The Alabama state legislature has little enthusiasm for the scheme. But in Mississippi, Georgia, and South Carolina, the plan is fully funded and ready to go, despite the protests of scientists and citizens.

A real danger from pesticides of all kinds is that they upset the balance of nature. The cotton bollworm, for instance, is kept under control by its natural enemy, which is a predatory insect. When DDT is sprayed on a cotton crop, it destroys the predatory insect but not, for some reason, the bollworm. The bollworm is therefore free to multiply rampantly. This kind of disturbance to the natural checks and balances has happened in numerous food chains throughout the world.

Still other pollution problems have been created by agriculture's use of artificial fertilizers. Many farmers concerned with maintaining the quality of their land rotate their crops. That is, they may plant corn in a field one year, soybeans the next year, alfalfa the next, and then leave the land fallow the fourth year. But some farmers put their land to intensive, constant use (in some places reaping more than one crop within a year). This has only been made possible by the use of artificial fertilizers.

Nitrogen is one of the key ingredients in most of today's fertilizers. Chemists learned some years ago how to take some of the world's free nitrogen and incorporate it into inorganic compounds which can be worked into the ground and utilized by plants. But nitrogen fixed into fertilizers that are used in steadily increasing amounts seems likely to unbalance the world's natural nitrogen cycle. When more nitrogen is put into the soil than can be released through the natural cycle, the surplus is carried away by water running off the land. The dissolved nitrogen drains into lakes and streams where it provides food for algae, causing them to multiply rapidly. The decomposition of this surplus vegetable matter uses up too much of the oxygen in the water. Gradually, a lake or slow-running stream begins to clog up with dead matter that cannot decompose properly after the water's supply of oxygen has been depleted. Plants and fish, which need oxygen to survive, die. If pollution continues, in time the lake or stream will not be able to support any living things. (In some places surplus fertilizers have even caused chemical pollution of drinking water.)

Agriculturalists also make modern farming more productive by feeding animals intensively in restricted feedlots. Under these conditions, so much manure is produced that it cannot be left to lie as natural fertilizer for the soil. Such large quantities overload and damage the land. It is a sad thing that the economics of farming now make manure one of the most unwanted wastes in the country because some farmers find it cheaper and more convenient to use artificial fertilizers. As a result, manure

31

Careless disposal of the vast quantities of manure produced on feedlots contributes to the pollution of our waterways.

is generally flushed away into sewers or allowed to run off into lakes and rivers where it causes serious contamination. In the United States, animals on confined lots produce a problem of sanitary waste disposal equivalent to that from 155 million people.

The agribusiness is gradually realizing the undesirable consequences of intensive agriculture and learning the dangers of introducing alien materials into the world's ecosystems. Much success has already been achieved by breeding more disease-resistant plants and animals and by developing special high-yield seeds. Attempts are also being made to control pests biologically — by breeding and setting loose great numbers of their predators. The predators themselves are first sterilized to prevent them from multiplying into a major problem. If modern agriculture can meet its problems in these ways, then the natural life cycles of the world will have a better chance to survive.

Mining

There is a mountain in Kentucky that is systematically being torn down, yard by yard, for the sake of the meager seams of coal which run through its rock strata. The area had been mined once already and the substantial seams removed. But now, with the aid of power shovels and earth-moving equipment, the mountain is being totally destroyed. Screaming machines and flying dust make the scene of destruction at this *open pit mine* even more hideous. The mining company employs armed guards to

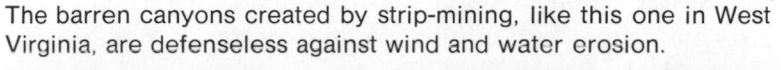

The barren canyons created by strip-mining, like this one in West Virginia, are defenseless against wind and water erosion.

keep out anyone who may wish to inquire into or protest against the rape of this land.

More than 20,000 *strip mines* are now gouging scars across the face of America at the rate of an additional 153,000 acres a year. One White House study estimates that by 1980 more than 5 million acres will have been defaced to feed our mining industries. Kentucky has lost 200,000 of its lush green acres. In Florida, 150,000 acres have been destroyed. In Minnesota, the U.S. Steel Corporation boasts a tourist attraction from what it bills as the biggest open pit mine in the world. Virginia, Colorado, New Mexico, South Dakota, Montana, and other states have suffered in similar ways.

One state, Kentucky, has now passed strict *reclamation* laws insisting that mining companies must rehabilitate the land they plunder by restoring the topsoil and by planting vegetation and trees. The law is beginning to take effect, but it will be many years, if ever, before the scars of the past can be erased.

Where the land is abandoned and bare without its topsoil, the process of building up soil by the actions of weathering, bacteria, and lichens starts all over again. But it takes thousands of years for the land to effect its own cure.

To some extent, the very vastness of America's land works against man's desire to rehabilitate what has been destroyed. The normal influences of supply and demand in a free enterprise economy mean that when land is plentiful, it is also cheap. When it is in short supply, it is expensive. In some of the more crowded European countries, land which is rehabilitated can be sold for a price that goes some way toward repaying the cost of rehabilitation. But where land is plentiful, as in so many mining areas of the United States, its selling price is low and may not cover the cost of reclaiming it.

An increasing number of people now believe that mining companies must include the cost of rehabilitating the land they destroy in the price at which they sell the minerals and metals they mine. This would raise the price of the raw materials of industry, but would put a stop to what many people think is the immoral practice of providing cheap raw materials at a high cost to the environment. To be effective and fair, strong reclamation laws need to be enacted and enforced equally throughout the country to prevent mining companies which are careless of the environment from undercutting the prices of other companies which restore the land they work.

Underground mining is also guilty of damaging the land. A recent government report recommended that legislation be passed to check the widespread damage to the environment now caused by underground mining. About 2 million acres of land surface have caved in over underground mines, nearly all of them over coal mines. Another 6 million acres are believed to be in danger. The report also detailed the damage caused by water pollution, by accumulation of mining waste into great piles, and by fires which break out underground and also in waste piles.

Slag heaps and *tailings* piles, which accumulate where ores are mined and processed, are very unstable, like mountains of loose soil with no vegetation to hold the soil in place. The wind whips dust from them over the surrounding countryside. When waterlogged, they can move in enormous landslides. Aberfan, a Welsh mining town, lost a whole generation of its children when a huge slag heap crashed down on the school soon after classes had begun one morning. The entire building was destroyed and most of the children were killed. For years the slag heap had sat tall, black, and menacing on a mountainside above the town. Then, undermined by a series of heavy rainstorms, it lost its footing and came sliding down.

Where uranium ores are processed to extract uranium for use in nuclear weapons and in atomic power plants, great piles of radioactive waste and tailings accumulate. The wastes are blown over the land and are carried away by rain, spreading radioactive substances over the countryside. As a result of contamination from such tailings, the Platte River in Nebraska has one of the highest radioactivity counts of any river in the United States. In Grand Junction, Colorado, efforts were made to find a use for uranium tailings that had accumulated there by mixing them with concrete used for building homes. However, when the United States Department of Health made tests on these buildings, they found levels of radioactivity considered too high for safety and the buildings had to be evacuated.

Even our national wilderness areas are not safe from profit-hungry miners. The Boundary Waters Canoe Area stretches for 200 miles along the Minnesota-Ontario border. It is a protected wilderness administered jointly by Canada and the United States. For six decades its defenders have fought off threats of exploiting the area for hydroelectric power, for timber, for minerals and furs, for resorts, and for landing areas for float planes. In 1969 over a million acres seemed to be protected under federal ownership. Then disaster threatened. George St. Clair,

When in 1966 a slag heap avalanched down the mountain into Aberfan, Wales, most of the children trapped in the school building (center right) lost their lives.

a New York City contractor, and Thomas Yawkey, the owner of the Boston Red Sox, announced that they intended to prospect for copper and nickel in the very heart of the Canoe Area and that they had the legal right to do this under an 1872 mining law. This law allows anyone holding mineral rights, even if they do not own the surface of the land, to enter any area for prospecting and mining.

The Izaak Walton League of America is taking legal action to protect the Canoe Area from such a violation. The League has called attention to other, more recent laws that conflict with the old mining law. A presidential committee at work on the Boundary Waters Canoe Area as long ago as 1953 advised: "If mineral deposits of major value are found, the public welfare must be the deciding factor in their use and development. If it cannot be demonstrated that their commercial use is of greater value than the wilderness that will be destroyed, such use should be prohibited."

The rival interests seem set for a long drag through the courts. The final judicial decision in this case may have a profound effect on the safety of other national wilderness areas faced with threats of destruction by private industrial interests.

Smelting

Many important metals of the world lie mixed with rocks and other minerals in underground storehouses. Miners extract ore, rocks that are especially rich in metals, from the ground. The valuable metals and minerals are then separated out by heating the ore to very high temperatures in a refining process called *smelting*. This process, unfortunately, gives off deadly fumes. And in a number of states, these fumes have turned the land around them into a desert.

The fumes contain large amounts of toxic substances, such as the gases sulfur dioxide and hydrogen fluoride, which kill off the vegetation that gives the land its protective cover. Without vegetation to hold the topsoil in place and to renew its fertility through decomposition, the land loses its ability to absorb water and to support life. The topsoil gets eroded away by the rain and the wind. After the topsoil is gone, the subsoil begins to erode. Streams dry up. The bare land suffers extremes of temperature. In time, the land becomes a lifeless desert where

This smelter in Tennessee had destroyed the vegetation over an area of 100 square miles before 1917 when it stopped polluting the land and air with poisonous fumes. In 1970 vegetation was only beginning to return to this man-made eroded desert.

nothing will grow. This is exactly what happened to the rich, woodland country around Copperhill, Tennessee, where smelters allowed all their poisonous fumes to escape from their smokestacks.

In Florida, south of Lake Okeechobee, smelting for phosphates produced such toxic fluoride gases that cattle were poisoned through their grazing feed. The teeth and the bones of the cattle were so damaged that they could no longer eat. They weakened and died. Now that part of Florida can no longer support cattle, and the ranches have had to be relocated.

In the Black Hills of South Dakota, near the turn of the century, a plant that was smelting gold ore gave off toxic and damaging gases. The company also poured out onto the ground thick, heavy waste residues from the refining process which damaged the land so severely that even 70 years later, nothing would grow there. The land may eventually recover—after a period of thousands of years—by going through the slow process of soil formation by the actions of frost, rain, wind, and bacteria. The only other way the land can be rehabilitated is by trucking in huge quantities of new topsoil and by replanting trees and groundcover. So far, no one has cared to make the investment necessary to bring back to life this land that has been destroyed.

Gradually, if the control of pollution from industrial smokestacks becomes more effective, we can hope that such destruction of our land will cease. But strong, effective controls are vital because more and more smelting and refining are likely to go on as we develop methods of recycling our waste products.

Most of the damaging chemicals pouring from smelters'

smokestacks can be held back by *electrostatic precipitators* which trap the pollutants by magnetizing them with an electrical charge. But precipitators are expensive to build and operate and, therefore, push up the price of the smelters' end products. In a competitive industry where the company selling at the lowest price is likely to get the biggest orders, smelters and other refineries have been reluctant to install expensive equipment to control the pollution they cause. Only strong antipollution laws, strictly enforced nationwide, will stop refineries from pouring their loads of filth onto our land.

Underground Contamination

Industrialists have been getting away with using brutally primitive methods to dispose of their manufacturing wastes, particularly in flushing them out into rivers, lakes, and the seas. But in the face of growing criticism from government and citizens, they are being forced to find other methods of disposal. One relatively new way to banish industrial wastes from the face of the earth is to sink them deep down into the ground in waste disposal wells. On first thought, this seems like a good idea. But investigation shows that this method, too, has its own special dangers.

Waste disposal wells are made by drilling a five-or six-inch hole deep into the ground until the bore of the drill reaches a layer of porous sandstone or limestone which is capable of absorbing huge quantities of liquid like a sponge. (This layer must be far below the level of any underground water supplies, and it may be 2,000 or 3,000 feet or more below the surface.) The drill hole is then encased in steel so that wastes cannot seep out on their way

41

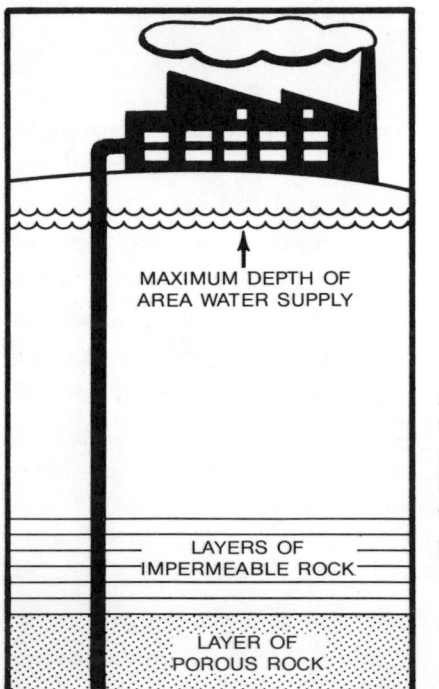

MAXIMUM DEPTH OF
AREA WATER SUPPLY

LAYERS OF
IMPERMEABLE ROCK

LAYER OF
POROUS ROCK

A waste disposal well. Liquid wastes are pumped under pressure down a pipe which leads to a layer of porous rock far below the level of any underground water supplies.

down the well and contaminate water supplies. When the well is complete, untreated liquid industrial wastes and sewage are pumped down it into the layer of porous rock. With luck, layers of impermeable rock will confine the wastes to the porous layer. But sometimes, when wastes are pumped into a well under high pressure, they fracture the impermeable rock strata and then leak out of the porous layer.

Ten years ago about 30 of these wells were in use. Today, official estimates put the figure at 150. But there may well be many more, some of them built before many states began to require permits for waste disposal wells and some of them drilled secretly and illegally. Such wells are cheap to drill and operate when their costs are compared with the costs of the continuous reprocessing and *recycling* of industrial wastes necessary to make them safe for disposal in other ways. Manufacturers are under constant pressure from their competitors and consumers to keep their prices down. It is easy to understand why they often jump at the chance of disposing of dangerous pollutants inexpensively, without inquiring too deeply into the long-term consequences.

The cap of a U.S. Steel waste disposal well at Gary, Indiana. The man holds in his right hand a sample of impermeable shale from 2,095 feet below ground. His left hand holds a sample of the layer of porous rock, about 4,000 feet below ground, into which waste sulfuric and hydrochloric acids are being pumped.

The U.S. Steel Corporation pumps hydrochloric and sulfuric acids from steel plants at Gary, Indiana, into this kind of underground dump. The Velsicol Chemical Corporation uses wells to dispose of poisonous residues from its manufacture of insecticides. Chemicals from Dupont's Teflon plant in West Virginia are pumped underground. The foul-smelling liquid waste from many of Hammermill's paper mills goes into disposal wells. So do the flammable liquids used in making textiles and plastics by Standard Oil of Ohio. All these wastes, and many more, are pumped down holes in the ground to where they are out of sight, and usually out of mind.

But some of them have unexpected ways of making their presence felt. At Ludington, Michigan, Dow Chemical has a well leading to a dump 2,500 feet below ground. Its steel lining corroded at only 450 feet below ground and leaked out thousands of gallons of brine into an underground freshwater supply. Fortunately, the water had not at that time been tapped for drinking supplies. But now it never can be.

In Denver, the underground pressure caused by the poison gas the Army had been pumping into wells there is believed to have triggered off a number of earthquakes in the area. One month after pumping began, Denver experienced its first earthquake in 80 years. During the next five years, there were 1,500 earthquakes, which varied in frequency in direct relation to the Army's periods of pumping. In 1966 the Army stopped using these wells, and the frequency of the earthquakes decreased.

In Erie, Pennsylvania, the Hammermill Paper Company forced its pulp waste into a disposal well at high pressure. One day trouble developed and the well blew, regurgitating the waste in a 20-foot high geyser that spilled into nearby Lake Erie. Waste continued to spout out for three weeks before the well could be capped. A similar blow-up of a well used by a citrus fruit processor happened in Orlando, Florida. For no known reason, the processor suddenly had in his plant a gushing geyser of diluted orange juice and other citrus wastes.

In Port Huron, Michigan, chemical wastes pumped into wells in the area by several manufacturers caused a build-up of underground pressures so that old, abandoned oil and gas wells started to leak. Crude oil pushed up through the concrete of a parking lot at the city post office. Oil seeped through the basement floor of a private home. Flammable natural gas spurted out just a few feet from the local hospital.

Only a few companies have made the detailed geological surveys that are essential if wastes are to be pumped underground without risking major disasters. Many industrialists are so thrilled to find a waste disposal method which appears not to harm the environment that they determine to go ahead and use it without inquiring too closely into what its long-term effects may be on the structure of the planet or its ecological balance. Without careful regulation of waste disposal wells, we could find ourselves living on the edges of volcanoes.

Waste disposal wells are not the only source of underground contamination. Every few months in the United States, massive explosions are deliberately set off underground to test improvements in atomic weapons. The Atomic Energy Commission has set off over 200 of these explosions, at least 17 of which have ruptured the ground or blown out their drill plugs and scattered radioactive materials into the atmosphere. In 1968 the Atomic Energy Commission's "Boxcar" test caused the earth's surface to subside, resulting in a crater 120 feet deep and 1,100 feet in diameter. In 1970 the "Handley" test was exploded deep within a remote desert mesa in Nevada, to test the warhead for an antimissile-missile system. The huge thermonuclear explosion created a cavern 800 to 1,100 feet in diameter beneath the earth. As the earth cooled off from the heat of the explosion, there was a massive collapse of the earth's crust into the cavern.

This huge crater was created in Nevada in 1968 by the Atomic Energy Commission's "Boxcar" test.

Some experimental work is also being done using nuclear explosions underground to release reservoirs of oil and natural gas trapped in rock strata. The "Rulison" shot was exploded in Colorado to release natural gas —which it did. But the gas was so contaminated with radioactivity by the nuclear explosion that it could not be used without endangering life. The Atomic Energy Commission's scientists tested the gas one year after the explosion and found it highly radioactive. But they said then that they expected to find it usable after one more year. However, other scientists believe that the gas will remain radioactive for many years.

Coal mines are yet another cause of underground contamination in many parts of the country. Miners work in the constant fear of cutting into pockets of explosive gas, which frequently lie close to seams of coal. When tragedy strikes with a mine explosion, there is often no way in which the fire it starts can be put out. When all hope of recovering the men trapped underground has been lost, a mine is usually sealed off and left to burn away underground. The coal seams burn slowly because of the limited supply of oxygen below ground. (In Pennsylvania, there are mines still burning 50 years or more after an explosion.) Sometimes the fire works its way up close to the surface, burns off the vegetation, and causes the land surface to cave in. A mine which has been closed down after its coal seams have been worked out can be dangerous too. After the mine is sealed off, the timbers supporting its tunnels begin to rot and give off an explosive gas. On occasion this gas has leaked through fissures in the earth and been set on fire by lightning.

Cultivation of Ugliness

In 1969 two men walked on the moon. That was the first time any creature from our planet Earth had set foot on the barren satellite. When the men lifted off from the moon to return to their own land, they left behind them several tons of litter, scattered around the area of their landing. Some of the litter consisted of scientific instruments designed to be used and then abandoned on the moon's surface—to remain there forever. Other litter included metal and plastic equipment, gravity boots and space suits, plastic bags of excretion, and other bits and pieces. As much as possible was jettisoned to lighten the lunar module for lift-off.

There are no living things on the moon to see all that junk, just as there may be nobody around most of the time to see beer cans sunk into wild mountain rivers. When worn-out farming equipment and vehicles were discarded in a remote Wisconsin meadow, no one looked on with horror. When a small town in Arizona dumps its garbage over the brow of the nearest hill, few people complain about the ugliness. But yesterday's remote meadow has become today's outlying suburb. Today's wild rivers will be the recreation areas of tomorrow. As the growing population spreads out over the vastness of America, the litter of earlier generations ends up in our own backyards.

Sadly enough, many people are so accustomed to seeing litter around that they are not offended by it. Surrounded by ugliness, they accept still more ugliness without protest. So scenic highways are lined with billboards—many of them excruciatingly ugly. Towns and countrysides are cluttered with the lines and cables and poles of power

A street in Seattle, Washington, hidden behind an ugly clutter of poles and power lines. The street was immensely improved when the lines were moved to a nearby alley in 1968.

49

companies. Whole sections of towns are hidden behind sign boards and flashing neon lights. Stores, homes, and public buildings by the million are constructed without a thought of how they will look, of what contribution they will make to "America the beautiful."

In a poor country still struggling for survival, some ugliness may be forgivable. But the richest country in the world surely should look better than it does. The trap is that much of America's wealth is generated by people who will cheerfully accept ugliness as the price of prosperity. Advertisers make money from their billboards. So if a tree gets in the way of a billboard, it is profitable to chop down the tree. Power companies can offer electricity more cheaply when they string their lines from poles, than when they have to spend money on burying them in the ground. And when electricity is cheap, people use more. When one businessman puts up a neon sign, a competitor wants a bigger one to compete. "Bargain" construction companies and their clients prefer to avoid the fees which good architects and designers charge.

In the years since World War II, about 6 million acres of countryside in the United States have been covered with little houses on their own little lots—most of them built in total disregard of architectural standards, urban planning, or the traffic problems they cause. Grass, trees, wetlands, and wildlife are destroyed to make room for them. Most cities have planning commissions, but usually their codes are not substantial or comprehensive enough to protect the environment or to prevent eyesore developments. They are chiefly concerned with checking proposals from real estate developers to see if they comply with zoning laws. As a result, towns and cities develop haphazardly, a block at a time. And most of them look as if they did, with just a few brilliant exceptions. Columbia, Maryland, is a handsome new city that was planned and built as a whole.

Countries less wealthy than ours are often more concerned with the beauty of their towns and cities. For example, in many of Europe's fine, old cities, the waterfront areas have been developed for all citizens and visitors to enjoy. Trees, gardens, and boulevards along the waterfronts delight the eye in Stockholm, Copenhagen, Paris, Zurich, Naples, and Florence. In the United States, except for a few cities such as Chicago and Washington, D.C., the waterfront areas are most often heavily lined with factories, wharves, and railroad tracks. Even in

San Francisco, California, surrounded by unimaginatively designed houses crowded together in long rows, provides examples of poorly planned housing developments.

suburban areas, the waterfronts are often inaccessible to the general public because they are lined with private homes.

We have all too few examples in the United States of how handsome a well-planned community can look. We need determined educational programs to help people in all walks of life to understand that cities can be beautiful, and that new developments, even of inexpensive housing, can be beautiful. Only if we learn to care and to tell our elected representatives in government that we care will the United States, the wealthiest country in the world, learn to build communities as handsome as some of those in poorer countries.

Many Americans believe that they have an inalienable right to do whatever they like with their own plot of land and that any movement to set up laws to control that right would be nothing less than a Communist plot. Yet there is a growing awareness that only planning involving some control over what the individual can do with his land can make America beautiful. Some of the complicated issues of long-term planning are discussed in Part 4 of this book.

4

The Garbage Problem

Disposing of Solid Waste

In America's early colonial days, city dwellers threw their household garbage into the streets and turned their hogs loose to feed upon it. This was a primitive form of recycling waste which few people would find acceptable today. But, even more to the point, hogs wouldn't find today's household garbage acceptable either. Old colonial rubbish was mostly food waste in one form or another. But the rubbish thrown out from modern American homes also consists of bottles, cans, plastics, cardboard, newspapers, old clothes, bedding, junked furniture, refrigerators, cars, and a million other "consumables" that are only partly consumed in our affluent lives.

Population growth, increasing affluence, technological developments, and unrelenting commercial enterprise are all at the root of the problem of waste disposal we face. American families are turning out waste in ever-increasing

amounts. The United States Public Health Service estimates that 20 years ago the average citizen discarded
about two pounds of trash a day. Now the amount is
nearer six pounds a day for each man, woman, and child
in the country. Where once people took shopping baskets
to market and carried unwrapped goods home in them,
today we arrive at the supermarket empty-handed and
leave laden with paper, cartons, jars, bottles, and cans
in a volume that may well be greater than the amount of
food we have bought. We throw away clothes and shoes
long before they are worn out because they are out of
fashion. We throw away carpets and furniture still good
for years of use because we can afford something new
which we like better. We change cars for the excitement
and prestige of a new model. We replace serviceable refrigerators and washing machines for the sake of a decora-
tor color. As it becomes increasingly expensive to get an
honest repair job done, we throw out faulty radios and
television sets and replace them with new ones.

All of the pressures of free enterprise suggest that our
habits of buying and discarding will not change signifi-
cantly. The United States maintains its great wealth by
constant stimulation of its consumer market. And it would
be unrealistic to suggest that any kind of control could be
imposed on the way individual families choose to spend
their money. But if we are going to live affluent lives, we
must face the problem and the expense of disposing of the
mountains of waste and garbage that result.

And private homes are certainly not the only sources
of refuse today. Municipalities, industry, commerce, and
agriculture add their share with sewage sludge, ashes and
cinders, chemicals, scrap metal, mineral tailings, radio-

active wastes, worn-out vehicles and machinery, materials from demolished buildings, agricultural crop and animal wastes, and much, much more. All of these wastes have to be disposed of somewhere and, year by year, as the volume grows, America's solid wastes threaten to engulf the nation.

The United States generates a total of about 3.5 billion tons of solid waste each year. This includes about 250 million tons of household, commercial, and municipal waste; 110 million tons of industrial trash; 1.1 billion tons

Children find the "fishing" good at the dumps within our cities.

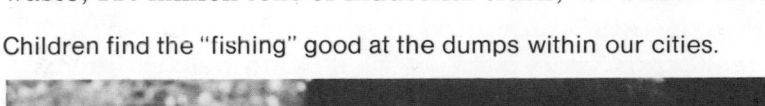

of mineral waste; 550 million tons of agricultural crop waste; and 1.5 billion tons of animal waste. Each year in the United States, our household refuse alone contains some 30 millions tons of paper, 4 million tons of plastics, 26 billion bottles and jars (over 135 per person), and 48 billion cans (over 250 per person). Where does it all go?

In many cases we just throw it on the land, where it creates health hazards and other dangers. The mining town of Lead lies in a wooded valley in the Black Hills of South Dakota. Lead still has only a few zoning laws, but when the town was started in 1870, there were none at all. For 60 years waste from the whole town was just dumped onto the side of a mountain about half a mile down the valley. Then one day, after a series of heavy rainstorms, the whole pile slipped and slithered all the way down the mountainside into the bottom of the valley. The accumulated garbage and trash of 60 years blocked the highway from Lead to Deadwood. It dammed up and further polluted a stream that already carried a burden of raw sewage from Lead and of cyanide and mercury from the Homestead gold mine. All of the garbage had to be scooped up, put into trucks, and hauled away to another valley where it was again dumped in the same stinking condition.

Despite rapid growth in the amount of solid waste and changes in its content, disposal methods have changed little over the last 50 years. One sanitation engineer commented dryly that the most recent improvement in waste disposal he could see was that waste was now hauled away in powered vehicles instead of in horse-drawn trucks. It is odd to think that we, the modern Americans who make hygienic palaces of our kitchens and bathrooms, just scatter over 80 percent of our garbage on the land we love.

Nearly three-fourths of the nation's garbage waste is disposed of in open dumps, which are a source of air, water, and land pollution.

About 73 percent of the nation's rubbish is trucked to open dumps, then spilled out onto the ground to be picked over by scavengers, dogs, children, and flies. Bulldozers push the rubbish into piles which are set on fire. The burning waste sends filthy plumes of soot and gases into the sky. At night, the rats come out, swap diseases with the fleas, and then spread the health hazard far beyond the boundaries of the dump. The stink is awful. Some pollutants carried by rainwater reach rivers and lakes. Other pollutants seep into the ground with rainwater and find their way to underground water supplies. This, in the space age, is how we treat nearly three-fourths of the nation's rubbish.

A much smaller amount, 15 percent, of our country's rubbish is burned in incinerators, many of them operated on a massive scale by municipalities. If the incinerators were thoroughly efficient and burned waste at tremendously high temperatures, they could reduce everything to harmless water vapor and carbon dioxide, plus fly ash which can be trapped in precipitators. But few of them work this way. Most burn waste incompletely and send up into the air soot, smoke, hydrocarbons, sulfur dioxide, hydrogen chloride, oxides of nitrogen, and the poisonous phosgene gas produced by burning plastics. The Public Health Service has found that 225 of the 300 municipal incinerators in operation are inadequate for reasons of health and pollution.

Eight percent of our rubbish goes into what is called "*sanitary landfill*," although the Public Health Service reports that of the 12,000 landfill sites surveyed in the United States, 94 percent are not sanitary at all and represent "disease potential, threat of pollution and land blight." Sanitary landfill is a disposal method in which garbage and refuse are dumped into open pits and each layer of waste is covered with a layer of soil. When skillfully designed and managed, and sited so that rainwater runoff cannot pollute any water supply, landfills can be a satisfactory way of disposing of solid wastes. But more often than not, this kind of landfill becomes unsanitary because too little soil is used for the amount of rubbish. Soil contains bacteria that decompose matter. It also contains oxygen needed by the bacteria for their work of decomposition. Without enough oxygen (because not enough soil), the bacteria are unable to break down garbage into humus, so each layer of garbage putrefies and builds up pockets of toxic, foul-smelling, and explosive gases. In the 1930s a group of archeologists were digging for artifacts in an area that had once been a dump outside the walls of ancient Rome. When they struck the 2,000-year-old garbage dump, the gases and odors were so vicious that the archeologists had to wear gas masks to continue their work. The results of inadequate landfill have given some similar nasty shocks to construction companies that have started digging into former landfill sites for the foundations of new buildings or highways.

A sanitary landfill. Trash and garbage are dumped into open pits, and each layer of waste is covered with a layer of soil.

When the ground is broken open, the stink causes distress for miles around.

But perhaps the greatest problem of all, even with the most sanitary landfill, is that cities all over the country are running out of space for disposal and, in their desperation, are driven to destroying much-needed recreation areas by filling in valleys, lakes, and bays. Chicago estimates that all its available areas for waste disposal will have been used up by 1975. New York City is gobbling up 200 acres a year for landfill and is fast running out of locations.

Few cities are willing to pay the cost of hauling their wastes hundreds of miles to parts of the country that have

Many auto junkyards are scattered along our nation's highways.

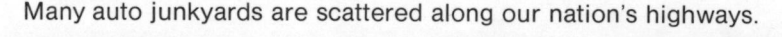

room to spare. And few areas will accept garbage from a far-away city to clutter up their own backyards. When New York City tried to make arrangements to take over an ugly, abandoned strip mine in Pennsylvania for sanitary landfill, the frenzied citizens of the area stampeded their state officials into vetoing the project. So New York still has to find space for its 6 million tons of refuse a year— enough to fill the Empire State Building 30 times over. And the gaping pits, some of them 200 feet deep, remain an eyesore in Pennsylvania.

Junked and abandoned vehicles present special disposal problems, and auto junkyards are another scar on the face of our land. They hold the rusting bodies of the cars Americans were once proud to own, which now lie piled up in huge, rickety mountains. We wish the cars would disappear, but they don't. Sometimes the vehicles are shredded and sold as scrap metal. Sometimes they are compressed into bales about three feet long and dumped into landfill. Sometimes they just lie on the land and rust.

The 1965 Highway Beautification Act was designed to control billboards and junkyards. It stipulated that by mid-1970, 17,000 junkyards were to be moved or screened. But the complex surveys, studies, and negotiations needed to complete this work were not done. Congress failed to appropriate the money which the act had authorized to compensate junkyard owners for complying with the law. So, by 1971, only 137 junkyards had been moved; only 1,518 had been screened. The number of cars in the United States multiplies even faster than the population. There are now about 7 million junked cars to dispose of each year. Yet still we seem not to care enough to put teeth into our laws.

We fool ourselves into thinking that when we have thrown something away, it ceases to exist. In truth, it is impossible to throw something away. What we are doing is throwing it into the air, into the sea, or onto the soil. Every year the amount of our solid wastes is increasing while the composition of those wastes is changing. And, because of our generally primitive methods of disposal, these wastes are polluting our land, air, and water. They carry the threat of disease. They are a desecration of the land.

Recycling Provides a Solution

By our actions, we have proved ourselves capable of setting a man on the moon, and yet so far we have been unable to solve the nation's garbage problems. There can be no city, state, or federal official who is not aware that problems exist, yet nowhere are there long-term, rational plans for solutions. And garbage and rubbish go on piling up. They have to be coped with every day, in ever-increasing quantities, taking up ever-more space on this land of ours. Our wastes won't go away while we sit and make long-term plans.

For a short-term solution, we can make greater efforts to reduce the bulk of waste by incinerating it. Efficient incinerators that do not pollute the air are just becoming available. But we can hope that strict enforcement of laws against pollution of the air will boost the development of new ones on the large scale needed by city sanitation departments. Such incinerators can even help men to utilize the energy that still remains in waste matter. For example, Hempstead, New York, uses heat from its refuse incinerator to run a 2,500-kilowatt electric power station and also a 420,000-gallon-a-day desalting plant, thus balancing the cost of waste disposal by putting the resources of the waste to another use.

Shredding, compacting, and baling provide us with another short-term solution by converting vast mountains of rubbish into more manageable proportions. One advantage of this is that hauling the rubbish long distances becomes more practicable. And if refuse can be hauled more easily, we might be able to fill in abandoned mines, quarries, and old sand pits formerly thought to be too far away to be useful. (San Francisco, which for years has been filling the edges of its beautiful bay with rubbish, is now considering hauling the wastes 300 miles away into the desert.) Another advantage is that landfill containing compacted rubbish is much more stable than that containing loose trash, to the extent that it can sometimes support buildings. Compacting, however, is suitable only for certain forms of rubbish. In Japan, it was discovered that the internal decay of tightly compacted garbage containing many organic materials produced dangerous gases.

Some cities have found inventive ways to utilize their wastes by converting landfill sites into commercial and recreational areas. New York City's La Guardia airport was built on sanitary landfill; so were large areas of Staten Island. When dressed with topsoil and planted with grass and trees, the garbage of yesterday has become the parks and golf courses of today. In Los Angeles, garbage was used to fill an open pit mine which has now been made into a beautiful botanical garden. Children in Detroit have a fine hill for sledding and skiing, built entirely of garbage and trash. But this kind of development costs a great deal of money. And sooner or later, we will run out of space.

Even worse, how crazy we are to spend so much money on hiding away valuable materials like glass, metals, wood, paper, and organic compost, when many of these materials could be recycled and used again or converted into some other useful commodities.

Glass bottles and jars can be crushed and melted down for use as new bottles and jars. They can be ground up and combined with asphalt for use as road paving. Experiments also suggest that they can be ground up and used instead of sand as the filler in concrete.

Paper, wood, and cardboard can be shredded and pulped and remade into a variety of paper products including cardboard containers, corrugated paper, wallboard, newsprint, hardboard, insulation material, wrapping paper, and paper toweling. In the United States, we use about 60 million tons of paper a year but recycle only 20 percent of it, even though we must chop down between 14 and 17 live trees to make one ton of paper. The United States Forest Service estimates that by the year 1985 the demand for wood and wood fiber will exceed the annual growth in our forests. Extensive recycling could help to satisfy this demand.

Cans and many kinds of scrap metal can be sorted and melted down so that they provide new supplies of copper, lead, iron, gold, aluminum, zinc, tin, and other metals. A research department of the Bureau of Mines found that they could recover metals to the value of $14 per ton from the ash residues in city incinerators. Selected trash can be as rich in gold and silver as some of the low-yield ores being mined now in Nevada because it contains thrown-away jewelry, coins, photographic negatives with silver

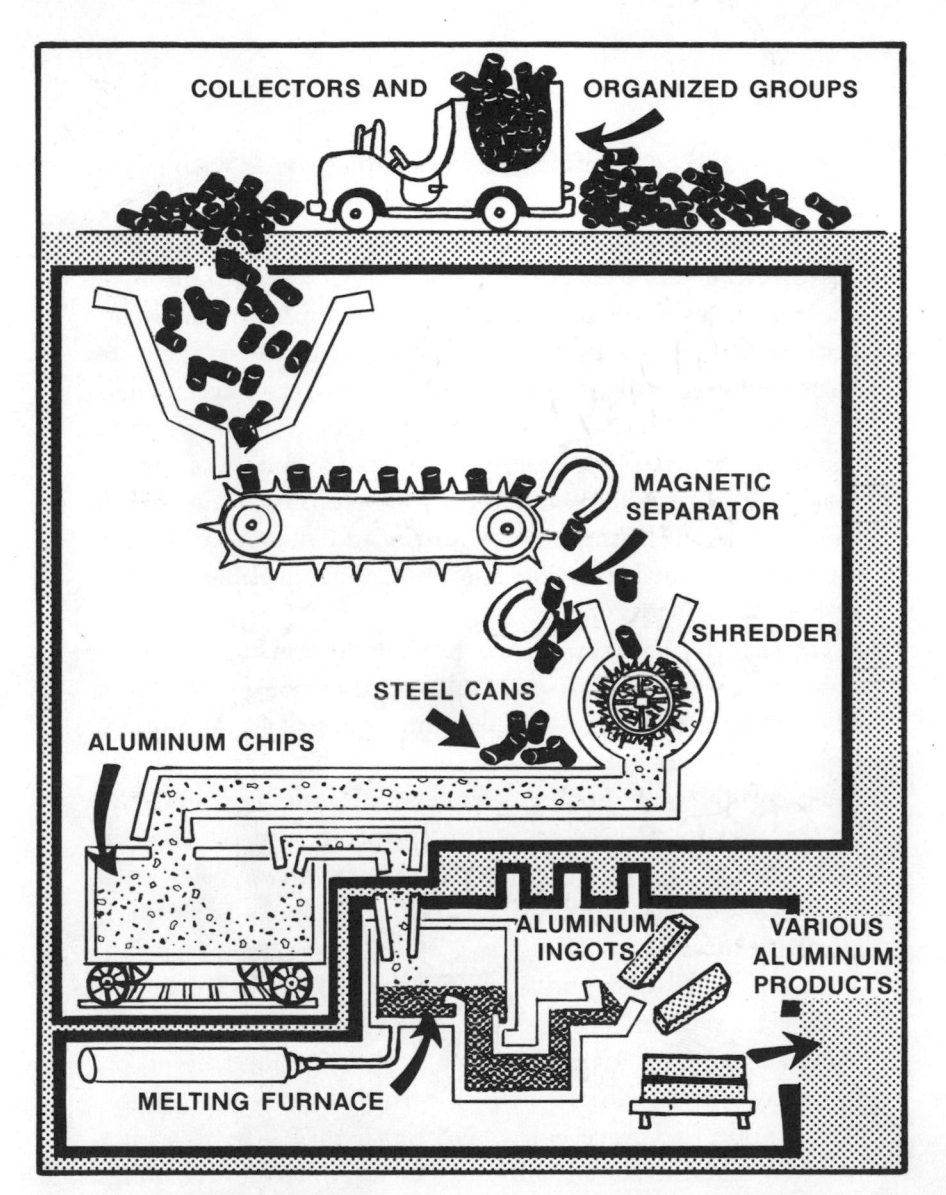

COLLECTORS AND ORGANIZED GROUPS

MAGNETIC SEPARATOR

SHREDDER

STEEL CANS

ALUMINUM CHIPS

ALUMINUM INGOTS

VARIOUS ALUMINUM PRODUCTS

MELTING FURNACE

How an all-aluminum beverage can is recycled. Cans are collected and the all-aluminum ones taken out by a magnetic separator. The aluminum cans are shredded into chips and then melted. The liquid metal is cooled and formed into ingots which can be used to make various aluminum products.

coatings, and commercial printing copies made with silver solutions.

Tires can be recycled as the various kinds of bumpers used in marinas and parking lots, as filler in buildings, and reprocessed for use as latex in certain kinds of paint. Indeed, it would be perfectly possible, if the incentive were strong enough, for automobile manufacturers to plan their production so that at the end of a vehicle's life, it could travel along an assembly line in reverse and have all of its reusable materials taken out for recycling. This would conserve natural mineral resources and also would go a long way toward meeting the enormous problem of disposing of junked vehicles.

Organic food waste from household garbage can be processed into compost and humus for enriching the land. The manure produced in enormous quantities on animal feedlots also makes excellent fertilizer. Even fly ash, the waste product of modern furnaces and incinerators, can be trapped by electrostatic precipitators and be used in building and highway construction. It mixes well with concrete and can be manufactured into blocks for sound-proofing buildings.

So why do we squander all these riches? Why do we bother to reclaim only a tiny 4 percent of the nation's solid waste? The simple answer is that in our free enterprise system we do not find it profitable to save them. Glass manufacturers can make more profit by working from raw materials than by collecting, sorting, crushing, and melting down old bottles and jars. (With present methods of garbage collection, hauling, and disposal, it costs more to dump a glass bottle than it does to make it in the first place. But if there were some way in which we could attach the cost of disposing of an item at the point of its manufacture, then we might, for instance, see rapid changes in the enthusiasm of the Glass Manufacturers' Association for throw-away bottles.) Paper mills, many of which have made heavy investments in their own woodlands, find it more profitable to fell and pulp live trees than to transport, sort, pulp, and control the quality of recycled paper. Farmers are unenthusiastic about organic compost and feedlot manure because they find chemical fertilizers both easier to use and cheaper. The government subsidizes mining with depletion allowances but does not subsidize processes of recycling metals. Most recycled metal products cannot be completed at a competitive price. Even freight rates are lower for iron ore and for pulpwood than they are for scrap metal and scrap paper. Because the economics of recycling is so discouraging, we have little incentive to search for more efficient methods of recycling that might lower the cost of the finished product.

In addition to these, another complex problem that must be faced by those interested in recycling is how to sort rubbish into its various materials for reprocessing. New methods of collecting and hauling garbage have to be found to replace the antiquated methods of today in which the many different components of waste are all mixed together.

The idea of recycling, however, is far from new. For years, scrap dealers collected and reprocessed many metals and large quantities of textiles and rubber. But the rising costs of collecting and separating waste materials caused the trade to stagnate. Metals that have a high value for their weight and volume or that can be separated

This giant pulper is one new aid for reclaiming waste. It turns truck-loads of unsorted municipal garbage into a watery mixture from which paper fibers, compacted metal containers, and glass are more easily salvageable.

out magnetically continue to be reprocessed. In the United States, about half of the copper, lead, and iron we use is still recycled, but only about 30 percent of the aluminum and 20 percent of the zinc. We reprocess less than 10 percent of textiles, rubber, and glass and 20 percent of paper.

Every industry has to meet competitive pressures to operate profitably. Yet, because industry does not find it economical to use much of the recyclable material available, the nation has to spend $4.5 billion on getting rid of its household, municipal, and commercial wastes alone. Added to this is the incalculable cost in damage to the environment caused by spilling out industrial trash into lakes and rivers and oceans, by scattering mineral tailings over the land, by flushing out animal wastes from feedlots and packing plants into our waterways, and by sending up into the air plumes of smoke and corrosive gases from burning solid wastes. Furthermore, when all our methods of solid waste disposal are tightly controlled to prevent dangers to health or pollution of the environment, the cost of disposal will be very much higher than it is now.

Somehow we have to find ways to make recycling materials more profitable so that industry has an incentive to keep on reusing the nation's resources. Today, the best way to provide that profit incentive seems to be through some form of government subsidies for recycling processes. A few faltering steps have been taken to set up such subsidy programs, but strong public pressure is needed to help the plans gather momentum.

People in some areas of the United States have already put their weight behind recycling and are directly pressuring manufacturers as well as state and national legislators. For example, communities in at least 25 states have set up laws banning the sale of non-returnable bottles. Some parks have already banned the sale of beverages in cans and bottles. Housewives, students, and Scout groups are setting up community services for returning bottles and cans to manufacturers. As concern for the ecology grows, manufacturers may have to make more efforts to recycle the materials they use just to keep the goodwill of their consumers.

However, the need to conserve our natural resources in itself should be an important incentive for recycling. Because of our highly developed economy, we in America,

These Boy Scouts are cleaning up a California beach in cooperation with a reclamation and anti-litter program in Los Angeles sponsored by the Reynolds Metals Company.

who represent only 6 percent of the world's population, use about 40 percent of the world's resources. If the under-developed countries of the world were to raise their standard of living to that enjoyed in the United States, the world could not possibly provide enough fuel, minerals, timber, water, or food for everyone. If our way of life is to have a future, we need to evolve a whole new philosophy and technology directed toward developing a closed system in which all the resources we use to support life and to produce the goods we need and enjoy are recycled back into industry or back into the natural life cycles of the ecology.

70

5

Learning Respect for the Land

Forests and Parks

Every American shares a magnificent inheritance of 771 million acres of federally owned land—just over one-third of the total area of the nation. Another 122 million acres are owned by cities, counties, and states. Indian reservations occupy 50 million more acres.

These public lands include some of the most beautiful and least ransacked areas in the nation. Three million big game animals depend upon them. Fish flourish in 1,500 miles of streams that flow through them. Green plants which grow on them produce much of our oxygen supply. Rain falls on their clean ground and percolates through the soil to replenish our underground water supplies. City dwellers can get away from the mental stress of their overcrowded lives to the peace and beauty these lands provide.

In order to survive with sanity, we desperately need the room to breathe, room for recreation, and opportunities for close contact with nature that these lands offer. And we have a responsibility to future generations to preserve them intact. But so far only 210 million acres of the federal lands are fully protected as national forests and national parks. The remaining acres are held by government agencies, including the Bureau of Reclamation, the Bureau of Land Management, the Army Corps of Engineers, and the various branches of the Armed Services.

Under the administration of these agencies, federal land is leased for grazing, military training and testing, mining, logging, canal building, and setting up electric substations. The waterways are dammed to provide flood control and hydroelectric power. A master plan of wise regulations rigidly enforced might be able to prevent commercial interests from destroying the land they lease. But no such master plan exists. As a result, our heritage of federally owned land is stripped and scarred by open pit mines. Trees are removed in such quantities that rainstorms cause soil erosion. Wild rivers are dammed so they can no longer run free and support creatures which thrive in fast-running water. Beautiful valleys are drowned in the reservoirs which back up from dams.

In 1970 a massive report of the Public Land Law Review Commission recommended opening up still more of the public lands to private, commercial, and industrial use. Conservation groups throughout the country have strongly criticized the Commission's recommendations on the grounds that the members were too much influenced by powerful lobbyists for the interests of big business. Strong and noisy opposition backed by voting power will be needed to protect the lands which belong to us all.

Every Friday evening, long lines of traffic on the

These forest lands in Olympic National Forest are being well managed—many are not. Block-cutting the Douglas fir helps new trees to establish themselves. Seedlings do not have to compete with mature trees for sunlight, nutrients, and water in the cleared areas.

highways leading out of our cities show what efforts city dwellers will make to get out into the countryside. Increasing wealth, mobility, and leisure time mean that more people can afford to get away from the cities than ever before. And a growing population means there are more people to get away than ever before. However, some of our wilderness areas are in danger of being spoiled by the very people who make such efforts to get out and enjoy them.

On the summit of 14,494-foot-high Mount Whitney, the second highest mountain in the nation, in California's

Sequoia National Park, trash containers overflow with paper, bottles, and cans. The park service regularly sends trains of pack animals up the steep trail to the summit and down again to pack out tons of trash. The park superintendent has said that the use of remote areas has increased phenomenally during the past five years and is now causing a litter problem which threatens to overwhelm his staff.

And litter is not the most serious problem park administrators must face. Forest areas have always stood in danger of fires started by lightning. But today they are also threatened with forest fires caused by human beings. A forest fire can start when a match or cigarette butt is thrown carelessly into dry brush, when sparks fly from

Yosemite National Park. Litter is a problem in recreation areas all over the nation.

camp and picnic fires, and when fires are not thoroughly extinguished by pouring water on them and raking out the damp ashes. Even a discarded bottle can start a fire. The glass focuses and magnifies the sun's rays to a point of intense heat which can set dry kindling on fire.

Recreational areas closest to our cities are those which suffer the hardest wear and tear. But during summer vacation months, recreation areas throughout the country are beginning to feel the ill effects of overcrowding and careless use. During the busiest weeks of summer, solid traffic jams on roads leading into Yellowstone and other great national parks create noise and air pollution, which destroy what should be a tranquil environment. To control the problems created by large crowds, park administrators have had to limit camping to designated areas only and to set up heavy fines for anyone caught dumping wastes. In the Badlands of South Dakota, areas containing rare cacti, plants, and fossils have had to be fenced off to protect them from marauding "nature lovers." In the Windcave National Park, tourists have stripped the caverns of their stalagmites and stalactites. Visitors entering the Petrified Forest in Arizona have to be warned that they risk severe penalties if they help themselves to even the smallest piece of petrified wood. In the Boundary Waters Canoe Area on the Canadian border, the use of motor boats and snowmobiles has had to be restricted. To assure places with freedom from noise, many forests and parks are setting aside wilderness areas which may be entered only on foot or on pack horses.

Because natural environments provide a wide variety of experiences which can help to heal the damage caused to us by the stresses of city life, more federal lands should be allocated for use as wilderness and park areas—despite the pressures from commercial interests. But we must also learn to take better care of the public lands we already enjoy.

Development of Virgin Land

For many years Alaska was known as "Seward's Folly" because William Seward, the Secretary of State in President Lincoln's administration, was primarily responsible for the United States' purchase of the territory of Alaska from Russia for $7.2 million. After the Yukon gold rush, which occurred near the turn of the century, Alaska seemed even more to be land that no one wanted. A few whaling, fishing, and trading communities grew up along the coast. But for the most part, native Aleuts and Eskimos continued their age-old lifestyles, making the most of their harsh environment with few interruptions from the modern world.

But little by little, Americans began to visit Alaska to exploit its wealth of natural resources or to enjoy its rugged, unspoiled scenery and delight in its abundance of wildlife for hunting and fishing. Then, slowly, Americans began to populate Alaska and make it their home. Here and there, some damage was done to the land, and the scars of that damage remain. Yet most of Alaska remained virgin territory. But recently, the ecology of the 49th state has been seriously threatened by the discovery of oil on its Arctic Coastal Plain—the northernmost part of the state, which lies within the Arctic Circle.

Of all the virgin land that Americans have settled, the tundra of Alaska's Arctic Plain is the most vulnerable and easily damaged. Because the climate of the region is cold year round, much of the land is locked in permanent frost. In the far north, some of the land is frozen to depths of 1,000 feet—farther south, to depths of a few feet. Above this layer of *permafrost* lies a thin covering of soil, which freezes and thaws with the change of seasons and anchors the mossy mat of tundra vegetation. If this arctic vegetation is destroyed, it takes from 30 to 50 years, even with careful planting, to get it growing again. One of the reasons for this is that Alaska's arctic lands have only a very brief, though intense, growing season during the summer thaw. Another is that the thin layer of arctic soil is not rich in humus, primarily because very few of the bacteria which decompose matter can operate in such intense cold. Wastes of all kinds hardly decompose at all in the arctic. An even more important reason is that when the ground cover is stripped away, the layer of permafrost begins to melt and then continuous erosion sets in.

The scars caused by bulldozers building roads across Alaska's arctic lands have never healed. And recently, the total length of these roads has been doubling in one year. In summer, when temperatures are above freezing, crews exploring for oil bulldoze the insulating layer of moss off the surface of the permafrost. As the road thaws and becomes muddy, they bulldoze off the mud to keep the road passable. So Alaska now has thousands of miles of muddy ditches consisting of exposed and melting permafrost. And instead of filling up with vegetation as they would in a temperate climate, these ditches freeze solid in the winter and erode away in the summer.

Oil exploration crews have also been responsible for other kinds of damage to the frozen lands. For example, when prospecting geologists abandon camps they have set

The Alaskan tundra has been badly scarred by bulldozed roads. The federal land damaged by this old road bed is now subject to continuous erosion.

up, they often pull out leaving behind them shacks and huge dumps of waste of various kinds, which the arctic weather preserves intact. (Wartime stores left on the Aleutian Islands had not decomposed at all after 25 years. There was hardly any rust on the metal. The paper had not broken down into its organic components.) Oil prospectors have also polluted the land with sewage and liquid waste because there are few bacteria to attack sewage sludge and the frozen ground cannot absorb, let alone clean, waste liquids.

However, the real danger to Alaska's ecology posed by the discovery of oil in the north is the proposed pipeline which the oil companies want to have built to carry oil overland from the Arctic Plain south to the warm-water port of Valdez. The pipeline, designed to carry up to 2 million barrels of oil a day, would be 800 miles long, with at least five pumping stations along its way. The problem is

A construction camp on the tundra near Prudhoe Bay, Alaska, pollutes the earth with its flooded garbage pit (right) and flooded domestic sewage pond (left). Soil is being heavily eroded (lower half) where the protective tundra vegetation has been stripped from the land.

that oil carried by the pipeline has to be heated to 150 degrees or more to keep it flowing along. In the arctic, this means that the pipe could not be buried or allowed to rest on the ground, except where it passed over solid rock, because it would thaw the permafrost and thereby dig itself into a deeper and deeper pit, and eventually fracture. Where the pipe passes over frozen tundra and gravel, it may have to be raised six or seven feet above the earth.

Conservationists protest that a six-foot-high barrier running hundreds of miles across Alaska would not only be a terrible eyesore but would also interfere with the migration habits of caribou and other big game animals. Such disturbance of natural life cycles in Alaska might have highly unfavorable consequences for much of the state's plant and animal life and, ultimately, for man. Another cause for conservationists' concern is that unless construction work were carried out with unprecedented care, the heavy machinery and construction workers' camps would leave unerasable scars on the countryside. Conservationists also point out that Alaska's many earthquakes make the pipeline even more of a potential hazard to the ecosystem. One break in the pipeline would dump at least 25 million gallons of oil onto the land. The Alaskan ecosystem could be severely damaged because wherever oil covered the frozen tundra, no life could exist for 100 years, or more.

Fortunately for Alaska, the discovery of oil there has coincided with the dawn of the age of ecology. Concern for the environment is growing. More and more people are beginning to realize the dangers of tampering with natural ecosystems without first making careful studies to determine the likely consequences. In accord with such concern,

Some oil has seeped onto the tundra from drilling operations. Alaska's frozen lands require hundreds of years to recover from oil pollution.

the Department of the Interior has prepared a list of detailed conditions which must be accepted by the Alaska pipeline company before it can begin building. Under these conditions, the company must accept liability for any damage it causes. The company must also guarantee its ability to make good any damage by paying a $50 million bond in advance. Other conditions established by the Department of the Interior relate to timber cutting, pollution, cleanup of oil spills, construction, fire prevention, use of explosives, protection of wildlife, use of pesticides, and damage to the ground.

There is little hope that the whole of Alaska can remain an unspoiled wilderness. But at least there is a possibility now that, if enough people voice their concern, the development of America's last frontier will be undertaken with knowledge, wisdom, and care.

Land Use and Zoning

There are now over 204 million people living on the 3.6 million square miles of the United States, or an average of about 56 people for every square mile. Thirty years ago there were only 44 people per square mile (not even considering the land areas of Hawaii and Alaska). At the turn of the century in the United States, there were only 25. However, by the standards of many other countries, we still have plenty of room. West Germany, for instance, has more than 600 people per square mile of its land. But our population has been steadily growing in the United States, as in the rest of the world, and our relatively high standard of living means that we make more demands than any other nation on our natural resources.

However, we have reached the point now where we are beginning to realize that each acre of our land is valuable, not just for housing, industry, and commerce, but also for sport, recreation, and beauty. In addition, we need undeveloped land to increase our understanding of nature and to maintain the balance of the natural life cycles which make our planet capable of sustaining life.

So far our land has been developed haphazardly. At the moment, about half, or 1.129 billion acres, is agricultural land. Forests cover 600 million acres. Cities take up 182 million acres. The military establishment occupies 30 million acres. Indian reservations take up 50 million acres. National parks and preserves lie on 101 million acres set aside for them. And public land owned by cities, counties, and states totals up to 122 million acres. No government has ever produced a master plan for deciding what proportions of the land should be reserved for what uses. Although the need for carefully planned development becomes obvious as available land space diminishes, it is extremely difficult for any group of people with conflicting interests to reach agreement on an enforceable plan for development.

An area of marshland, for instance, means different things to different people. Those who live near it might not value the land at all because of the mosquitos which breed there. Real estate developers might like to fill the marshland to provide prime building sites. People living in overcrowded tenements might welcome the opportunity to move to a trailer park situated on the reclaimed marshland. Commercial fishermen treasure the marsh in its natural state because they know it is a rich breeding and feeding ground for the fish on which their livelihood depends. City authorities might think of the marsh as a convenient, out-of-the-way site for dumping their city's wastes. Nature lovers value the marsh for the richness and varieties of its wildlife. The list could go on indefinitely. To decide which use of the marshland is the best would take the wisdom of Solomon, based on a full knowledge of how much the conflicting needs are met in other areas and what effects changing the marshland would have on the ecosystem of the area.

An area of marshland being managed for wildlife. Two spaces have been cut out of the cattails to provide courting birds with open water. The hay bales give the birds a resting place. Lands like these are often developed—unwisely, in many cases—to fulfill private, commercial, and industrial needs.

The nearest thing we have to a master plan for wise use of our land is a mass of zoning laws established and administered by local authorities. These laws are intended to protect public health, safety, order, convenience, and general welfare. In theory, zoning laws should decide conflicts over land use, stabilize property values, prevent overcrowding, promote circulation of traffic, and ensure the harmonious development of the whole nation.

However, even if every zoning board in the nation carried out these high principles, using the finest judgment and taking into account the needs of future generations, we might still end up without a sensible national plan for land utilization. Local zoning boards have little knowledge of what is happening in the rest of the country or of what effect their local decisions will have on the ecology as a whole.

The truth is, of course, that zoning boards are humanly fallible. They can be pressured by the business interests of their communities. They feel they have a responsibility to encourage local industrial development that will provide jobs and prosperity for their electors. They have been known to spend millions of dollars on land development plans which they lack the determination or the power to enforce. Their laws are not enforced uniformly; often they make exceptions for individuals or corporations and grant spot zoning, special use permits, and variances.

In addition, for better or for worse, the decisions of local zoning boards can be overridden by higher authorities such as the county, state, or federal government. A city zoning board, for instance, which decided not to grant permission for building a new airport on a certain site, could have its decision reversed by the state planning commission. On the other hand, an area designated by a local board for industrial use could be reallocated by the state or federal government for park land.

At every level of government, there are many different departments, each primarily concerned with its own responsibilities and each subject to various outside influences. How a planning commission decides to use certain land may be determined by which interested department or outside pressure group at its level exercises the most power. But what is right for the housing and development authority may be wrong for the port authority. The needs of the highway department are likely to be in direct conflict with those of the conservation department. The area of land desired by an organized group of businessmen for industrial development may be just as strongly desired by another organized group of citizens for a recreation area. The winner of the resulting conflict may well be the group which shouts the loudest or shows the greatest power.

What happens when individuals or small groups stand in the way of more powerful interests? Public utilities have rights of eminent domain, which means they can take land through legal condemnation procedures to lay their gas lines, telephone cables, sewage pipes, or power lines. One farmer violently objected to an electrical power company's decision to erect huge towers and power lines on his land. He threatened to attack as trespassers everyone who set foot on his land to begin the work. He was arrested, put in jail, and let out only after the power lines had been erected. Such cases are bizarre, and everyone will agree that many times individual rights should yield to large-group rights. But individuals or groups who hold great power should not always be able to get conflicts over land utilization settled in their favor—just because they hold the most power. Planning commissions which have the strength and the wisdom to enact and enforce comprehensive programs for land use, without being totally controlled by powerful interests, are needed at every level of government.

Today city planning commissions need special strength and wisdom. The latest census figures show that Americans are continuing the trend of leaving the country and moving to the city. Many planning experts believe that the move to the city will continue until the nation's population becomes concentrated in three vast megalopolises: Bogwash, spanning the east coast from Boston to Washington; Frangelego, running down the west coast from San Francisco to San Diego; and Chicpitts, straddling the murky industrial areas between Chicago and Pittsburgh.

Such concentration might have the advantage of freeing other parts of the country from the threat of urban development. But will life be worth living in the big city areas under such highly crowded conditions? How can we maintain our sanity when we are constantly surrounded by concrete? Where can space be found for highways to accommodate weekend flights to open country? Is highly urbanized living as we now know it what we want? Do we know what we want? City planners are under extreme, constant pressure, not only to find more space to house more people, but to answer these larger questions and to solve the problems of their cities before they become unsolvable.

Far-sighted urban planners are calling now for national commitment to adventurous schemes to transform dying cities into pleasant places to work and live. They believe the technology needed to accomplish this is no more complicated than that needed to place men on the moon.

One such scheme proposes that streets, on which cars, buses, monorails, and other forms of transportation would operate, be covered with a concrete ceiling. The level above the ceiling would be open to the sky and lined with stores, businesses, and residential buildings. More important, it would be planted with grass and trees and flowers and provided with meandering paths where people could talk, read, relax, and amble. Buildings would be faced with terraces to hold ponds, trees, shrubs, and

One proposal for an ideal city of the future suggests that street-level traffic be covered by a roof planted with grass, trees, and flowers.

gardens. Such an environment would not only be very pleasant, it would also be healthier because the numerous plants, using sunlight, air, water, and soil, would be manufacturing clean, fresh oxygen. In the hearts of our cities, we could be in contact with nature.

Of course, to transform a present-day urban area into such an ideal environment would cost a great deal of money. But we have the necessary wealth and technology to accomplish such a project; we need the necessary commitment and wisdom.

Of the several plans and techniques developed in recent years to help us improve our utilization of the land by

Some planning efforts have already resulted in improvements of our urban environments. In Oakland, California, a roof garden open to the public was created above a garage and adjacent shop building.

assisting planning commissions at all levels, perhaps one of the most fascinating and comprehensive was originated by Ian McHarg, a professor at the University of Pennsylvania and a partner in a company of planning consultants. McHarg has developed a technique, using satellite and aerial photography and map-reading scanners and computers, by which planners can weigh up the many conflicting interests at war over a proposed development and be able to form a balanced judgment.

Suppose, for instance, that a new airport is needed — or a highway, golf course, or industrial park. Using McHarg's technique, planners would prepare transparent maps of all the possible locations for the development. Then they would draw up transparencies for each of the natural, economic, and human values to be considered. They would probably make separate transparencies to show the effects developing a particular location would have on wildlife, ground water supplies, land erosion, flood control, wetlands, recreation areas, and scenery. Still other transparencies would show such things as the importance of each location's distance from the metropolitan area to be

Ian McHarg's planning technique was used to determine "the area of lowest social cost" for routing highway I-95 between the Delaware and Raritan Rivers in New Jersey. Transparent maps were prepared to show where the highway construction would involve the greatest (dark areas) and least (light areas) damage in terms of various natural, social, and economic values. The top three transparencies show the areas of greatest and least social cost in relation, respectively, to susceptibility to erosion, urbanization, and recreational values. The bottom map is a composite of 10 transparencies prepared for 10 separate values to be considered. The dark arrows show the routing for the highway recommended by this study.

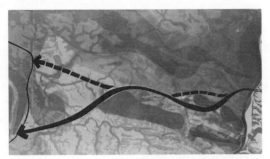

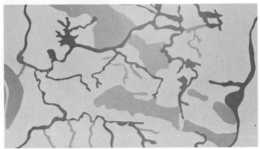

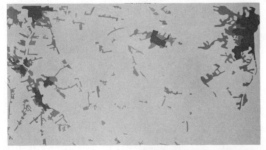

served, the immediate economic cost, and the ultimate economic cost of the project at each area proposed. These transparencies are made using graduated color so that the areas where a project would involve greatest damage or highest cost appear dark, and those involving least damage or lowest cost appear light. After all of the transparencies are ready, planners would then place them on top of each other on a lighted table. The area which showed up lightest would indicate the location for the project which represents what McHarg calls "the area of lowest social cost."

In an interview published in the *National Civic Review*, Ian McHarg explained: "My proposal is based not on sentiment but on society's own self-interest. When man puts houses with septic tanks on the recharge areas of an underground reservoir—so that he literally defecates in his own wells, as is happening all over the country right now—it's pure economic inefficiency and it shows up in higher taxes for water purification and in higher hospital bills. The fact is, there is a best place for everything you want to build, whether it's a steel mill or a ski resort. We should be searching the country for these places." McHarg has also called for a nationwide "ecological inventory" so that we will know what we have in the way of natural resources and land development. Once we know what we have, surely we can do a better job of planning intelligently for the future use of our land.

Needless to say, to benefit fully from techniques like McHarg's and from greater knowledge about our land, we need much more coordination among planning boards from different levels of government than we have now. And even with better coordination among planning

These houses were built on unstable land that a soil survey could have revealed was unsuitable for building development. The damage to the houses resulted from an earthslide. Better planning and stricter zoning laws could have prevented this disaster.

boards, there is still the problem of how many limitations the American people will accept on what many believe to be highly important freedoms. If a man wants to build his home or his business on a known earthquake fault, should he be free to do so? If he does build and then later loses his building in an earthquake, should he qualify for compensation from federal disaster funds to rebuild in the same place? Do people have the right to build on the flood plain of a river, where they risk losing everything in the spring floods and polluting the floodwaters with their debris? Who should have the power to prevent a man from building a beach cottage on fragile sand dunes when by building it he destroys the natural, protective pattern of the seashore, making it possible for storms to break up the shoreline. If someone chooses to ignore the danger of landslides and builds a house on the side of a steep canyon, who should stop him? Deciding how to answer these and similar questions will not be easy.

Prosperity or Survival

We know that human beings depend on green plants to provide them with food and with oxygen to breathe. We must plan to be sure our land can always support enough green plants to guarantee the necessary supply of food and oxygen for future generations. We know that land scarred by strip-mining is destroyed by erosion so that nothing more will grow on it. We must plan so that miners guarantee that the land will be reclaimed after they have taken out its minerals. At the moment only 12 states attempt at all to control reclamation of mining land. (Only in Kentucky are effective reclamation laws beginning to insure that land will be restored.) In some places, consumers are beginning to apply pressure on mining companies. For example, the Tennessee Valley Authority, a publicly owned power company, now purchases its coal only from suppliers who reclaim the land they mine. But these efforts are not enough. We know that the land is mother to all living things. We know the damage man is doing to the land and how this damage is disturbing the natural life cycles of the world. We must act strongly now if we are to survive.

A few years ago *ecology* was a word used only by specialists. Now it is in most people's vocabulary. It is even beginning to crop up in political speeches. But we need to be on our guard against believing that a string of high-sounding platitudes means that we are really meeting the problems of saving our environment. A politician who tells us he believes that land should be preserved and developed for human, natural, and economic values has said nothing at all unless he goes on to describe some of the specific actions he proposes. In fact, human, natural, and economic values are often incompatible. In the past, economic values have usually been given priority. To get our ecosystem back in balance, we must now begin to give priority to natural values.

There will be some serious consequences of giving priority to natural values. If our land is to be preserved, we must accept a slow-down in our economic growth. To accept this, we must first clearly realize the futility of creating economic prosperity in an environment which is no longer fit to live in. Now that we have learned to dominate and change the natural ecosystem, we have a special responsibility to understand the complicated balances of nature which make the ecosystem work, and to make sure that we preserve these balances.

The complete change in philosophy which is needed to reverse our values could take many years to develop. By then we may have so plundered our land that we are too late to save it. So every step we can take now to make wise decisions about the use of our land is vitally important— to the quality of the land we bequeath to our descendents, to the survival of the worldwide ecosystem of which we are a part, to our own survival.

Glossary

biological amplification. The process in nature by which a substance, such as the pesticide DDT, is distributed throughout an ecosystem, accumulating in greater and greater concentrations as it is passed up a food chain from one living thing to another.

consumers. All of the animals in an ecosystem. They are called consumers because they feed on fruits, plants, or other animals.

decomposers. Living things in a ecosystem, such as bacteria and fungi, which break down dead or waste matter into nutrients that plants can use.

ecosystem. The total system of relationships in a given area among plants and animals and their nonliving environment, which includes such things as climate, soil, and water.

electrostatic precipitator. A device to remove suspended particles of matter from a gas (as the gas released from smelting or incinerating). The particles are given an electrical charge and then the gas is passed through a magnetic field which collects the charged particles.

food chain. The pathway by which food and energy are moved through an ecosystem. The chain consists of living things linked together—each organism feeding on the one below it and being eaten in turn by the one above it in the chain.

irrigation.   Supplying dry land with the water necessary for plant growth through man-made canals, ditches, pipes, or floods.

open pit mine.   An open mine with a series of connected step-like ledges forming a road around its sides on which trains or trucks can carry ore directly from where it is being mined.

permafrost.   The permanently frozen layer of subsoil in arctic regions which varies in depth from a few feet to hundreds of feet.

producers.   All of the trees and other green plants in an ecosystem. They are called producers because they manufacture food for animals.

reclamation.   Restoration of land that has been damaged by mining, agriculture, industry, or anything else to a condition in which it can once again support plant and animal life and defend itself against erosion.

recycling.   Passing used materials through a series of treatments so that they can be reused, used in making new materials, or disposed of with safety.

sanitary landfill.   A method of waste disposal in which succeeding layers of trash and garbage are covered with layers of soil.

smelting.   The process by which a valuable metal or mineral is separated from its ore by heating the ore to great temperatures in a specially built furnace.

strip mine. An open mine created by removing shallow layers of soil and rock from the earth with power shovels to uncover seams of valuable minerals, especially coal, lying close to the surface of the ground.

subsoil. The layer of soil below the topsoil. Subsoil is coarser in texture than topsoil and is usually not fertile enough to support plant life.

tailings. The refuse material separated out from valuable minerals when ores are mined and processed.

topsoil. The layer of fertile surface soil, usually only a few inches deep. Topsoil is one of the essential parts of all land-based ecosystems because it provides the nutrients plants use and pass up the food chain to man and the other animals.

weathering. Any of the chemical or physical actions of weather which cause rocks to break down into soil.

Index

About the Authors

Pollution: The Land We Live On is one of eight books on pollution written by Claire Jones, Steve J. Gadler, and Paul H. Engstrom. This volume was a cooperative effort, each person contributing his or her own knowledge and experience, with the final result a kind of "literary synergism."

Paul H. Engstrom is a minister, a lawyer, and a family counselor, as well as president and cofounder of the Minnesota Environmental Control Citizens' Association. Under his leadership, MECCA has worked for preservation of Lake Superior and the Mississippi watershed, reduction of radioactive pollution, reuse of materials in solid waste, and many other environmental goals to improve the quality of life. Thus Rev. Engstrom's major contribution to this series of books on pollution was a social and legal perspective resulting from direct experience.

Steve J. Gadler also is experienced in the fight to save the environment; he is a registered professional engineer who was an environmentalist long before pollution became a national issue. A retired Air Force Colonel, Mr. Gadler has for many years been asking pertinent, revealing questions about the damage caused by our industrial society. He has been especially concerned about radioactivity, which is an invisible but deadly threat to life itself. In 1967, the governor of Minnesota appointed him as a member of the state's Pollution Control Agency. Mr. Gadler's technical expertise is apparent in each book in the series.

Claire Jones is an experienced writer who first became aware of the dangers of pollution in 1956, when she lived through one of the famous London killer smogs. Teaming up with Rev. Engstrom and Mr. Gadler gave her an excellent way to express her concern over the condition of the environment. However, her contribution has been more than a concerned citizen's point of view and a crisp, sparkling writing style. A native of England, Mrs. Jones brings a special international outlook to this series. None of the problems of pollution can be seen as less than worldwide, and this important perspective gives *The Land We Live On* added value.